# LIVE ONE

## 8 ELEMENTS TO EXCEL, ENGINEER & EMPOWER YOUR LIFE

### RICKY VENTERS P.E.

www.selfpublishn30days.com

*Published by Self Publish -N- 30 Days*

**Copyright 2017 Ricky Venters Enterprise LLC.**

Printed in the United States of America
ISBN: 978-1544611785

1. STEM 2. Success-Psychological Aspects

**Ricky Venters Enterprise LLC. : LIVE ON E**

**Disclaimer/Warning:**

This book is intended for lecture and entertainment purposes only. The author or publisher does not guarantee that anyone following these steps will be a successful leader. The author and publisher shall have neither liability responsibility to anyone with respect to any loss or damage cause, or alleged to be caused, directly or indirectly by the information contained in this book.

## This book is dedicated to:

*My Parents, Ricky & Robin for paving the way and the many words of wisdom along the journey.*

*My Wife and best friend, Nicole for your continued love and support.*

*My Sons; Solomon, Levi, and Kaleb my source of inspiration.*

*My Brother Ravon Venters my rock and motivation.*

*My Sisters, Wyatashia, Brittany and Rashawn.*

*My Grandparents, Joe & Annale Humpfrey, Alonzo & Delores Foster.*

*My Godparents: Earl & Donna McKinney*

*My Mother in law, Joyce Paul.*

*All of my Uncles, Aunts and Cousins:*

*Michael Venters, Angela Parks, Louie Parks, Jenifer Humpfrey, Basil Foster, Darrell Barnes, Donald Bellamy, Lestina Bellamy & all of my Family.*

*The Genas Family, Wright Family and Busby Family.*

## *My Accountability Brothers:*

*Donavon Genas, Marlon Madden, Jude Alfred, Curtis Vialva, Vaughn Edmeade, Eric Thomas, Mason West, Jeremy Anderson, Quest Green, Hugh Sinclair, Ian Ferguson, Maurice Wagoner, Lajuan Wideman, Sullivan Phillips, Andrew Nangle, Clayton John, Leighton Newell, Phil Simmons, Will Mack, Daniel Dawson, Jared Taylor, Johnwood Raphael, Peter Campbell, Shawn Arthur, Sean Thomas, Brian Thomas, Darren Palmer, Tarrence Motley, Roy Parham Jr. & Joshua Matthews.*

Special Thanks to the RVE & InVenters Team


Special Thanks to the Beacon Light Tabernacle
Family, Bishop Jesse & Jacquette Bottoms & The
Beulah Baptist Church Family, Pastor Paul and
Windy Graham & Restoration Praise Center Family.


Special Thanks to my Mentors & Colleagues:
Philip C. Steiner, Kari Nystrom, Adam Trojanows-
ki, Jeff Leavenworth, Craig Razza,
John Kohler, Rory Ronan, Joseph Lembo,
Steve Lembo, Peter Beltz, Robert Hedman,
Erik Bodelson, John O'Connell, Walter Yarmich,
Victor Alves, Ryan O'Conner, Merrick Hylton, Don-
ald Steiner, Doug & Jeanne Tebera,
Scott Dixon, Howard Reel, Donna Brinkmeyer,
Nancy Martin, Bobby Miller, Deborah Holifield,
Danielle Karavedas, Adam Smith
and Amy Mercurio.


Special Thanks to all of my friends that
supported me on this journey.

# RICKY VENTERS P.E.
www.rickyventers.com

## BOOK RICKY FOR YOUR NEXT EVENT!

DYNAMIC SPEAKER

PROFESSIONAL ENGINEER

CORPORATE LEADER

STEM ADVOCATE

# Contents

# INTRODUCTION

It was just another typical night. The excitement of beginning the school year was still pumping through my body. That excitement that has you eager to go bed early so that the next day could quickly come. To make it even better, I had just experienced my eighth birthday. Life couldn't get better.

As I went through my usual nightly routine, I remember my head spinning with excitement from all the plans for the next day. The teams for the basketball game during recess were picked, I had my favorite egg salad sandwich ready for lunch and my tradable items as well (oatmeal cookies to be exact.)

As I said my prayers and said good night to my mother and father, I laid down to sleep, but my mind kept racing. What will tomorrow be like, how much fun will I have, will John bring my favorite snack to trade, will my team win the basketball game? I guess we will see I said as my eyes finally closed for a good night's sleep.

Hours later I woke up crying for help. I

thought to myself, "No, this can't be happening again!" I woke up gasping for air. It felt like someone was reaching inside of my body and grabbing both of my lungs and squeezing as tight as they could. No, not again. I remember screaming, "Mommy, Daddy, God somebody help me!" My door burst wide open, and there stood my parents. They rushed to my bedside and held me tight. I remember my mother on the phone calling for help. I wasn't quite sure who she was calling but what seemed like seconds later, I heard sirens in the distance. Were they coming for me? Next thing I remember, three strangers rushed into my room. They put me on a stretcher and carried me out of the house and placed me into the ambulance. It all seemed to happen so fast. I remember dosing in and out of sleep and just repeating "Lord, Lord!" aloud. While drifting in and out of consciousness, I captured small snippets of time. The paramedics ripped my new sweatshirt. The X-Ray Technician was rolling me over on the X-ray table. Being wheeled down a hospital corridor. Each moment flashed by quickly leaving me no time to process what was happening.

"He's awake!" I heard my mother say. I tried to respond, but my body was so weak I couldn't even muster the strength to open my eyes all the way. I laid there motionless. It felt as though a heavy weight rested on my body restricting me from the slightest movements. I could hear the beeping of the monitors and wheeling of carts in the adjacent rooms and with my eyes partially opened I could see the blurred figures of my father, mother, and grandparents sitting around my bedside. Then suddenly an overwhelming peace came over me. I tried to recall the events that lead to this point but, the last thing I could remember was crying out at the top of my lungs "Lord help me!"

The next day I remember waking up with enough energy to open my eyes and see where I was. I had a brief moment of panic when I woke up and saw tubes in which seemed like every part of my body. I laid there motionless thinking wow how did this happen? How did I go from being ok to near death? The doctors told my parents that both of my lungs had collapsed and they had to put me on the ventilator system. I didn't have the lung capacity to breathe on my own until my body got stronger.

I was released from the hospital after three weeks with instructions to take my time and let my body regain its strength. So running outside and playing wasn't an option. Talk about torture. At eight years old all I wanted to do was play. Well, I had something else that was ready to entertain me, homework.

After returning to school, I was greeted with

open arms of love and a sheet of paper with all my missing assignments listed. What a greeting! So instead of running outside and playing, I spent my recess time catching up. It was overwhelming, and it felt as though I would never catch up. Well unfortunately, instead of the list shrinking, it grew long and longer by the weeks. Don't get me wrong, I wasn't incapable of doing the work, I just didn't care to do it. That's how third grade went up until the parent teacher conference.

When my mom returned from the parent-teacher conference, she walked in the house called me into the kitchen and said some words that would change the trajectory of my life forever. She said "Your teacher told me that you would never be good in mathematics. My career options will have little to nothing to do with math." Those words penetrated my whole body it was as though the breath I fought to keep a few months before was just punched out of me and my body went immediately into survival mode.

Those words left a scar in my mind and from that day forth I purposed to prove to myself and my teacher I can and will succeed in mathematics. I will not let her words place a limitation on my life. From that point on, I was determined I will be the best person I can be and strive for greatness.

With this book and associated workbook, I will share eight principles, which I call elements

I learned over the years to engineering my life to success. Proceed with extreme caution: if these elements are applied to your life, you will experience a life full of joy, happiness, purpose and contentment.

# *Attention to Details*

*Filter the unnecessary, distractions
aren't to be entertained.*

*Narrow the task ahead until
it's straightforward and plain.*

*Procrastination prevents precision.*

*It forces on the spot decisions.*

*You can't execute the vision if you never
took the time to simply listen.*

*Attention to detail comes from preparation
and organization.*

*I am specifically addressing my work with
clear communication.*

*My work represents me, shown with
etiquette as I am Elite*

*Always neat, I perform A-Class, not Class-C.*

**—IAN FERGUSON**

# ATTENTION TO DETAILS

Attention to Details - { $\mathscr{D}$ }.
The ability to concentrate and focus on information including insignificant material and high-level information.

As I walked into my house one summer day, I was overwhelmed with excitement from my first day at summer camp. I stepped on what was supposed to be soft carpet but instead my foot met a rough, rugged wooden surface. I quickly looked down and saw all the carpet was gone.

Confused as to what was going on, I asked my father who was busy pulling up the carpet in different areas of the house what he was planning on doing. Were we getting new carpet? Maybe my mother didn't like the current color? Maybe it was the juice I spilled a couple of weeks ago I thought I got it all up without anyone noticing? My Dad just simply said we are putting down wood flooring. Oh, I said in surprise. Why would we do that?

My mother walked in the house by that time after pulling the car in the garage after picking us up from camp and was very pleased with the progress my Dad had made while she was gone. Hearing my question, she responded the carpets hold a lot of dust. I quickly reflected on the poor job I did vacuuming the carpet at times and thought wow I was just learning you couldn't give me a little more time to perfect it. I didn't think I was doing that bad of a job. My thoughts were interrupted as my mother continued. Dust is one of the things that triggers your asthma so if we remove anything that could hold significant amounts of it from the house; we could control your condition better.

So every asthmatic should be able to identify what causes their symptoms. Those symptoms could be anything from; humidity, hot weather, dairy products, pollen, dust, laughter, weather change, physical exercises, crying, having a cold, large meals, stress, or grass and the list goes on. It's up to the individual to pay close attention to what sets them off, also known as an asthma attack.

For me, at some point, every last one of those listed has affected me in some way. I would find myself laughing too hard at a joke my friends shared and the next thing I know, I'm coughing followed by a breathing treatment. I would go outside after the grass had been freshly cut and after rolling around and playing for a few minutes, I'm inside because my breathing has been set off. I remember times when I've eaten so much that I can't breathe regularly. So you're reading this thinking how in the world does this guy live? Well, it's simple, I had to learn how to Live on E.

You know when you're driving a car, and your gas light comes on indicating it's almost out of gas and you are now on full alert seeking to find the next gas station? That panic that you feel, thinking I hope I have enough to make it to the next exit to fill up? Or when your cell phone battery is at two percent, and you are in a frantic search to find your charger just hoping the call doesn't drop before you do. That is what Living on E, or you can say "Empty" looks like. Every day on high alert, strategically monitoring your body and hoping that you can stay away from the triggers that will set you off.

No, you may not have asthma, but there is something in your life that is preventing you from functioning at your highest level. You have some goals in your life you can never seem to reach. Some relationships haven't blossomed the way you dreamed. Your passion for you job has minimized to just receiving a paycheck. There is something holding you back, and you have never addressed the root of the problem.

The first element that I will like to introduce is Attention to Details - { $\mathscr{D}$ }. The ability to concentrate and focus on information including insignificant material and high-level information.

In a society where everything is so fast-paced, and many of us suffer from the desire to have everything at our fingertips, this element of life often gets ignored. How often do you analyze your life? In other words, how often to you reflect on your journey? The good, bad, accomplishments, dreams, relationships or challenges? How often do you think about the things that bring you joy or those things that bring you pain? I never understood the statement the older you get; the faster time goes by until I began to experience it in my life. It seemed as though one day we are celebrating the New Year's and in a couple of weeks we are celebrating Christmas again. Scientifically, this made no sense to me. Based on this theory, two individuals that work in adjacent offices, pending on their respective ages, could finish their 8 hour work day earlier than the other even if they started at the same time. This idea wouldn't be an impossibility either because the older employee would arrive at the office earlier due to the accelerated time of their day. Confusing right? Of course, it doesn't make any sense.

Anyways, back to my point, since we live our lives at such a fast pace, we frequently forget to take a moment to pause and reflect on the life we are living. I remember when I was a young engineer designing the mechanical systems for various types of buildings. I designed system layouts for hospitals, schools, museums and corporate office buildings. In the design process, we always have someone review our drawings to assure the quality of the plans. Without fail, no matter how good the designer was, the reviewer found at least one thing that was missed or more clarification needed to be provided. Every time! I tried my best to produce a set of drawings that were perfect and needed no revisions but, regardless of my efforts, changes were sitting on my desk the next day. What I realized in this process was that sometimes to see the details of the plan, you may need to take a step back. I remember my boss telling me to put away a set of drawings I had been working on for a few days and to work on something else. He said, "You can't see the forest for the trees." I was so intimate with the project I was overlooking critical details.

Growing up I was so worried about controlling my asthma when I was having an attack, but I never took the time to identify the triggers that put me in that situation. I went day after day managing and treating the symptoms, but not realizing I would never be able to control my condition until I analyzed the details of triggers. I had the best devices to treat the attack, yet I failed to change the environment that fostered the attacks to flare up.

Ripping up the carpet was step one. My parents realized they had to modify the environment. Carpet is the most common flooring a house has due to the comfort it brings to your bare feet as you bury your toes in it walking across each room every morning. No need for slippers or socks. My parents were so determined to provide an environment that prevented these attacks that they were willing to compromise their level of comfort. How willing are you to compromise your level of comfort to prevent goals and dreams in your life from being shattered?

More time is spent on creating the goals vs. time on the establishment of the environment to accomplish the goals. Paying attention to details places you in a position to not only look at what you want to achieve but, also how to set yourself up to accomplish it.

Write out your top five goals and then list the environment you need to foster to achieve those goals.

**List your goals and the associated environments that will promote success.**

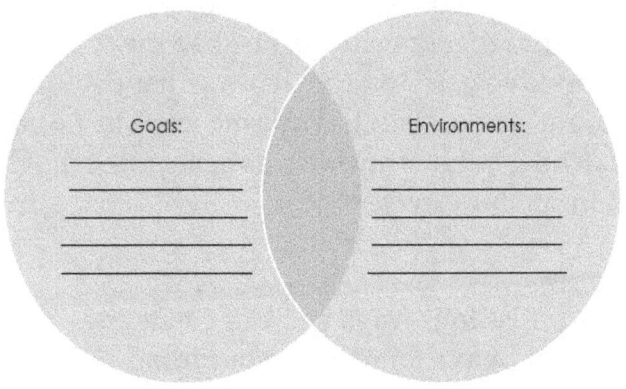

But what if you can't control the environment that you are in? What if I told you that your environment is not only physical but mental? The things you place in your mind shapes what you become. What do I mean? I remember a show called 24. I used to watch that show nonstop on Netflix. Season 1 to Season 8 and start back from the beginning again. The show was great! I loved every minute of it, but some things started to change. I noticed when I drove; I had a sense of paranoia like someone was always following me. I would stare into my rear-view mirror, observe everything and take notes of each car behind me. Then I noticed out of nowhere my interest for working with the FBI started to increase. Mind you, it was a faint thought in the past but, I never actually considered it. Here I was in a successful engineering career, but all I could think about was "man it would be great being out there catching some bad guys." When I started to catch on to what was happening all I could do was laugh. With the mere fact of me watching a show, my life was being altered right in front of my eyes.

**List four ways you can create the environment of your mind.**

_____

_____

_____

_____

## Reflection

Attention to detail gives you the ability to not only analyze the world around you but, also self-assess who you are. By putting this element to practice every day, you will be able to identify the things that will push you to greatness and also the things that will prevent you from obtaining it.

## Workbook Exercise

Now that you understand why paying attention to details is a vital element to engineering a life of success. Complete exercise #1 in the workbook.

Download workbook at
www.rickyventers.com

# *Communicator*

When I'm speaking to you,
I'm sharing what's already inside me.

Sharing my ideas, thoughts, dreams,
so that you might see

the value I see in you oh ever so proudly.

I want you to see it for yourself,
look at your life's location.

That's why it's so important for me to share this optimism with constant communication.

What good is the knowledge inside me
if it can't be applied?

To those I encounter, setting no boundaries
I want to speak life worldwide.

Internationally, across the globe,
changing the world around me

The ripple effect begins now; the pebble is dropped,
my major impact starts NOW.

**—IAN FERGUSON**

# COMMUNICATOR

---

Good Communicator – [ $\mathscr{C}$ ]
The ability to transfer your knowledge
of subject matter in a way that can be received
by any audience.

---

Everyone communicates in some form or fashion and over the years technology has increased our capabilities. I remember when I was an undergrad student at Stony Brook University and my then girlfriend now wife Nicole, who was studying at SUNY Albany, would communicate all of the time. We would write letters, AIM (AOL instant message), or call anytime after 9:00 pm when unlimited free minutes would begin. Constant communicating almost erased the 3.5-hour distance between us.

Now, this is not a book about relationships or a romance novel but, I would say the ability to have those lines of communication available is what kept our relationship strong throughout those four years. I remember the times when I wouldn't hear from her for a day, and I would go into panic mode thinking of the worst things that could've happened. Yeah, I was that guy, drunk in love and couldn't get enough.

During those time I would say what was most important was the quality of the conversations. Being separated for months on end, I had to become an excellent communicator. I needed to have the ability to connect with Nicole to keep her interest, or after a matter of time, someone else would've taken my place. I wasn't about to let that happen.

Everyone communicates, but not everyone is a good communicator. How many conversations can you remember when someone is trying to explain something to you or tell you a story, but you stood there lost and tried to figure out a way to change the subject or end the conversation altogether? We've all been in those awkward moments and find ourselves just nodding our heads as though we are truly engaged in the conversation. I remember a time when I was on opposite side of that conversation trying hard to deliver my point but couldn't seem to catch the right words to describe my thoughts. It went a little like this:

I was promoted to the project manager position and had my first opportunity to represent my company at an owner's meeting. I met my supervisor in the parking lot of the building and, as we walked inside, he said to me, "Ricky, you are going to be running this meeting. I'm just going to sit back and be there to support you." At first, I thought, "Great! Here's my shot to show what I can do!" Then, all of a sudden, nervousness hit me. As I was rehearsing in my head, I couldn't think of any words to use to explain my progress of the project. Then I heard what seemed like dreadful words, "Do the engineers have anything to add or any comments?" I started talking, and I stumbled way too much about issues no one at that table cared about. It was a mess! Finally, I said the best words of the meeting "That's all."

Feeling ashamed of my less than stellar performance, I tried so hard not to make eye contact afterward. As my supervisor and I walked out, I confessed, "I was so nervous." He looked at me laughed a little and said: "I didn't know what in the world you were talking about in there." Then he followed up with simple words that just seemed to stick to me, "Ricky, next time, just tell them what you know. Nothing more, nothing less." What he said wasn't profound, but it allowed me to realize that I didn't have to put on a show or try to look like the smart guy in the room. All I had to do was share what I knew.

What makes a great communicator? There are three types of communication; written, verbal, and nonverbal. We use each of these forms of communication every day at work, school or communicating with a friend. In today's time, written communication is done through e-mail, texting, or instant messaging. I remember a time when phone services would charge a fee for texting, so it was very limited, but slowly over time texting has become the primary source of written communication. I can go through months where I'll send over 5,000 text messages, crazy right? Verbal communication is the communication where a sound is transmitted and received through speaking and singing. Nonverbal communication is communications through body language, facial expressions, dancing, and movement.

We use each of these forms of communication every day, but the question still lies what makes an excellent communicator?

Dale Carnegie is known as the best communicator of all time. Why? Because he knows how to do one thing extremely well and, that's listening. We often mistake being a good communicator with one who is delivering a speech or can write exceptionally well, but don't necessarily consider listening as the major factor. I often found myself thinking about how I was going to respond to someone as they were talking to me; therefore, missing the essence of what they are trying to say. For some reason, we feel an overwhelming pressure to have something to say immediately after someone makes a statement or asks a question. It's okay to pause to gather your thoughts after fully hearing the statement or question and then responding. I guarantee your response will be more concise and make a bigger impact.

## Reflection

Being a great communicator can open doors to opportunities that you would have never imagined and, just like each of the other elements can be mastered if you practice and study great communicators. Without proper communication, a message can be received incorrectly that could lead to embarrassing and humiliating moments. How many times has autocorrect taken over your phone and the message you thought you were typing made absolutely no sense when received by the recipient? I think autocorrect has gotten the best of all of us and that's a perfect example of failed communication. In your mind, you know exactly what you're saying but somewhere in the translation your audience, friend, or a colleague is entirely lost.

Here are some tips to improve your communication skills.

1. **Practice what you're going to say.**

   If you're preparing to give a speech, take the time to write out what you're going to say and rehearse it over and over even to the point of memorizing it. Repetition is the main ingredient for perfection.

2. **Don't Rush**

   Take your time when speaking. I was once told to "Speak like a Queen" meaning take the time to pause in between sentences. It may seem very awkward at first, but it works.

3. **Listen, Listen, Listen**

   Don't feel the need to talk all the time; sometimes it's just best to listen.

4. **Body Language**

   Your body language should match your words. If someone told you, they were having a fantastic day, but their body language expressed sadness, you would be very confused. Therefore, make sure your words and body are in sync, or you will give mixed messages.

## 5. Take a course

Taking a course on effective communication or joining an organization like Toastmasters would take your communication skills to the next level. In those settings, they will give you opportunities to work on the different elements of communications we discussed.

**Workbook Exercise**
Now that you understand why communication
is a vital element to engineering a life of success,
complete exercise #2 in the workbook.

Download workbook at
www.rickyventers.com

# Critical
# Thinking

*Not to be confused with overthinking,*
*I'm talking strength in the mind.*

*Execution of thoughts in the brain*
*that is used in real time.*

*You have to absorb and unwind and*
*rewind the information given.*

*Don't let the same thoughts keep spinning time*
*to make real life decisions.*

*Use it at work; it's an element to keep all around*
*a skill to always use.*

*Not to abuse, but use to serve others.*

*Analytical thinking can be beneficial at any moment.*

*Pay attention, and observe,*
*learn and be wise with your words.*

*Wise like Solomon think critically to solve problems*
*and give answers never heard.*

## —IAN FERGUSON

# CRITICAL THINKING

Critical thinking [ *&#9998;* ]
The thought process of actively and skillfully conceptualizing, analyzing, and evaluating information gathered from, observation, experience, reflection, reasoning, or communication, as a guide to action.

One of my all-time favorite TV shows, Mac-Gyver, displayed the highest level of critical thinking a person can possess. C'mon I know it was only TV, but Mac could skillfully analyze any situation and find a way of escape. The show became so predictable and it seemed like the greatest challenge for the writer of the show was to place Mac in the most challenging and unlikely positions of escape. Mac was a pure example of a critical thinker through keen observation, and the ability to pull from his previous experience. He was able to develop a guide to action that would free him or anyone else needing to be saved.

I remember my early years of designing mechanical systems for multiple types of buildings and being in awe of the senior level engineers and partners. It seemed like they could create a system in a matter of minutes compared to my days. They would quickly go from building section to building section, identify the problem areas, and ways the problems could be solved. I tried to figure out what I thought the pattern was, applied it to my projects, and to my surprise, it worked! Well, I figured it was working. I got to the point I could fill the plans with an array of subtle details. The plan looked perfect until my drawings were submitted for review. I'll never forget this day. My same mentor I mentioned in the previous chapter looked over my drawings, puzzled, and began asking me questions. I was stumped and had no right answers to what seemed like straightforward questions. At that moment I realized that I didn't put any critical thought into my design. All I did was mimic what I saw others do and though I perfected emulating them, my final product failed.

I wish I could write a book sharing all of my highlight moments in life to make me look like I had it all together, but rather I need to show you my pitfalls and shortcomings and how I made the adjustments. I had to make a decision that day. How was I going to move forward? Was I going to continue trying to take short cuts or go through the process of dissecting each component of the project and understand it like the back of my hand? I decided I needed to go all in. This was one of the hardest decisions, and the change didn't happen overnight. Slowly, over time, I was able to develop myself to the point of reviewing other designer's projects.

As a millennium, one danger we often face is not utilizing our critical thinking ability. Honestly before reading the definition, how many of you could've defined what critical thinking is? We live in an era where things are thought out for us already. For example, how many phone numbers do you have memorized or how many DOS codes do you use to launch your computer programs? Is this a bad thing? absolutely not! but it supports my argument that though the technology we love so much has made life easier for us, at the same time, it limits the use of our mental thinking. Nowadays critical thinking is used only in response to a dangerous situation; it's not something that's practiced every day. We don't even use critical thinking while driving anymore. On long trips, my parents would take out a huge atlas and map out the route we were going to take. Fascinated by all the highways and streets shown I would just read through the atlas and find the connections of roads to surrounding cities. Over time technology has made it easier for us. Instead of carrying a map in the car, all you have to do is go to MapQuest on your computer and enter your destination, and in 2 seconds your whole trip was charted, and all you have to do is press print. Then we progressed to the TomTom or the Garmin, whichever one you preferred. Everyone was raving over these and had to have one in their car because instead of having an atlas or printed directions from MapQuest you had a GPS system in your vehicle that called out the instructions as you drove. No more thinking, just listening. Now we have progressed so much that the GPS system is embedded directly into the cars or cell phones and once we turn it on, it gives us suggested destinations based on your travel history.

When I leave my house in the morning Monday through Friday, my GPS automatically directs me to work, and likewise, when I leave work in the evenings, the GPS leads me home. It's so sophisticated that it will give me real-time traffic reports and send me on the shortest route. It's amazing how much technology has enhanced our way of life, but here's the problem.

One day while traveling home my GPS started taking me in a direction I never went before, but what I didn't realize was that my phone was on 1% battery life. So of course as you could've already guessed, my phone dies meaning my GPS is gone; therefore, I'm lost. I was on back roads, with nowhere to get directions. The two things I had going well for me was that I had a full tank of gas and it was summer. Why did it matter it was summer? If you haven't guessed it by now, the answer is because the sun was still up. If this would've happened in the winter, it would've been dark before I even left the office. Ok, what difference does it make if the sun is up, you may be asking? Well, one answer is it's always great to be able to see where you are going and the area surrounding you, especially on back roads. The other answer I'll explain like this; when I was a young kid, I used to be fascinated by the correlation of the position of the sun and the time of day. Our house was positioned in a way where the side of the house was facing north and south, and the front and back were facing east and west. Throughout the day, while I was playing outside, I was able to keep track of time by only looking at the position of the sun. So now I'm in this situation of being completely lost. The first thing that came to mind was knowing my house is located north of my job. If I keep heading in that direction, I will eventually end up on a familiar road. I quickly looked at the time in the car and found where the sun was positioned. I knew at that time of day the sun would be west so after locating the sun I used it as my guide to point me in the north direction.

Finally, I made it to a road I was familiar with, and I was home bound. Critical thinking in full effect. God forbid anyone that swears by their GPS be placed in that situation you can only image the stress that would arise.

Now I didn't say all of this to say technology needs to be banned and  shouldn't advance. It should raise awareness that even though it seems to make life easier, don't necessarily mean it makes us better.

## Reflection

As the years pass by people that have the ability to be critical thinkers are going to be far and few between. If you take the time now to enhance your critical thinking skills you will stand heads and shoulders above your competition in the field you're currently in or decide to pursue.

**Process to increase your critical thinking**

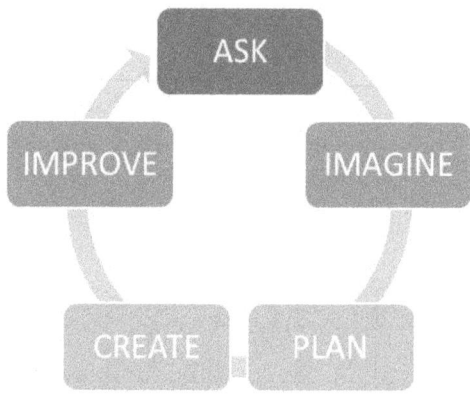

**ASK:** Identify the problem.
**IMAGINE:** Identify all possible solutions.
**PLAN:** Develop a guide to action.
**CREATE:** Carry out your plan of action.
**IMPROVE:** Find ways to improve on your ideas.
Start the process all over again.

It was through critical thinking that an Atlas transformed into an app on your cellphone or how songs playing on a record can now be played on your cell phone. The birth of new technology is a result of someone asking a question, imagining possible solutions, putting the plan into action and finding ways to improve.

## Workbook Exercise

Now that you understand why critical thinking is a vital element to engineering a life of success. Complete exercise #3 in the workbook.

Download workbook at
www.rickyventers.com

# *Persistent*

*Fast forward the actions its takes
to meet my goals for sure*

*Keep pushing with persistence, non-stop,
til' I can't go anymore*

*If effort is all it takes, I'm confident I won't fall short.*

*I'm a train, can't be aborted, my concentration stubborn,
stone hard like a fort*

*Pursuing my passion with consistency*

*Day after day, I'm hungry for this,
no one knows how much it means*

*To become what I'm destined to be, starving for success*

*I can sense it*

*Smells so good I can taste in in my dreams*

## —IAN FERGUSON

# PERSISTENT

Persistent [ $\mathscr{P}$ ]
The ability to continue the pursuit of a goal
despite any obstacles that arise.

I remember it like it was yesterday, my oldest son was about four months old and laying on his stomach during "Tummy-Time" as my wife called it. For the longest time, I thought this was a name she made up, but I was eventually proven wrong. My wife and I often stared in amazement at this little human we created, and as we stood there, my son began to move back and forth. I stood there silently wanting to see what he was going to do next.

Then it happened, he flipped over! We shouted and cheered as though he won his first award at school. He laid there with a huge grin on his face as though saying "just wait till you see my next trick." A couple of months later he was able to pull himself up and cruise around the living room. He stretched his arms for dear life as he went from one object to the next. Nothing was safe anymore! Everything had to be put away, out of his reach. Then what seemed like days later he began to stand on his own without any support.

Finally, the day was here. The day one of the most cherished milestones took place. During one of his routine cruises around the living room, he spotted a toy that was out of his reach. Instead of falling to the ground and crawling over to it, it became a moment of bravery. You could see it in his eyes, the determination to get to that toy on his feet. He took the first step towards it with his fingers still grasping the edge of the couch. He went to take the second step but, realized after this step he would be all on his own. To step forward, he had to let go of the support behind him. He took one look around and then went for it. His legs shook, and then he fell down on his pamper padded behind. I thought he was done and was just going to scoot or crawl the rest of the way but, to my surprise, he got back up and took another step. Smack was the sound made when his bottom hit the ground again. I thought for sure this was it but, again to my surprise he got back up and did it all over again. After a series of tries, he finally reached his most prized toy, and he reached it standing on both feet. After that day, walking became the norm and never again did he shuffle his body across the floor.

As I recapped that moment, a flood of questions came to my mind.

1.  What gives a baby the desire to walk?
2.  How many times is a baby willing to fall before walking comfortably?
3.  Why do they never go back to crawling again?

**What gives a baby the desire to walk?**
Studies show that one main contributing factor to babies walking are Primitive Reflexes (PR). These reflexes are the first part of a baby's brain development and remain active for the first few months to a year. Without going too deep into the science, as a part of these reflects kicking and stepping motions as displayed.

### How many times is a baby willing to fall before walking comfortably?

After understanding one main reason why babies have the desire to walk, I think the answer to that question could be summarized as such. In a baby's mind, it's not the amount of times they fall, that's counted, but rather it's a burning desire inside of them to explore the limitation of their abilities.

### Why do they never go back to crawling again?

They see the world from an entirely different perspective when they transition from crawling to walking. Everything is more accessible, they finally move around like everyone around them, and there is a feeling of pride and joy by reaching this milestone in life.

### Have you seen it yet? Have you seen why persistence is one of the eight elements?
### Let's take a look at your life and ask yourself three questions:

1. How will persistence benefit me?
2. What happens to my instinctive nature of persistence?
3. How do I get persistence back?

### How will persistence benefit me?

Persistence led Thomas Edison to create the electric light bulb and Lewis Latimer to invent the carbon filaments that allowed the bulbs to last longer, to be more efficient and cheaper. Persistence pushed Gerald Lawson to develop the Modern Home – Video Gaming Console that led to the birth of Xbox, PlayStation, and Wii. It was through persistence that Dr. Maria Telkes and Eleanor Raymond built the first home entirely heated by solar power and that Hugo Taran Salguero invented the Rotary Engine.

Persistence is the tool that always beats adversity. It is inevitable that difficult times will come into your life. There is no way to pinpoint when they will come or how devastating it will be, but one thing for sure, we must be ready when they arrive. Without persistence, it will be so much easier to quit in life. There would be nothing to push us beyond the resistance of achievement. No great feat as ever been accomplished without it. You can't afford not to be persistent. Like the baby, you can't afford not to explore the limitations of your abilities, anything less would lead to a life full of unfulfilled dreams.

### What happened to my instinctive nature of persistence?

Well only you can ultimately answer this question, but I can suggest to you somewhere along life's path you become comfortable and complacent. Persistence can't reside with complacency.

They are total opposites, and in this case, opposites do not attract. There are very few people in this world that enjoy struggling or pain, and sometimes solitude in complacency helps shelter them from it, for a season at least. As with any muscle in your body, if you don't use it, it becomes weaker, and that's exactly what's happening to your instinctive nature of persistence.

### How do I get persistence back?
Here's one tip: place yourself around people that have unyielding work ethics. Surround yourself with people that don't give up when difficulties arise and are high achievers. Here's the thing; one reason why babies have a desire to walk is due to the simple fact that everyone around them is walking. The power of positive association will push you to new heights and pull things out of you that you would've never dreamed. The rebirth of your persistence lies in the desire to reach your full potential.

## Reflection

Persistence aka "Drive" is what separates winners from losers and I know it sounds cliché, but I've seen it displayed over and over in my life. My father would always tell me "Never give up. Never quit." Those words always stuck with me, and when I activated that persistence, I began to grow professionally. There was a time in my life where nothing in my life was changing. I remember having the desire to become better, but I didn't have the drive to stay up late at night working or studying. I remember a time when I wished I had certifications and licenses where I would be recognized as someone serious about my profession. I would write the initials of the desired achievements behind my name in pen hoping that would activate a fire inside, but It wasn't until I read and applied the context of this e-mail did my drive wake up.

> To: "Ricky Venters Jr."
> Date: Thursday, August 25, 2011 7:51:29 PM GMT-0400
> Subject: Re: Looking for some Advice
>
> I think what you put in your mind and who you surround yourself with is crucial. I wake up and pray every morning for about an hour, keep my goals in front on me, and I only run with ppl who are doing the same and who will hold me accountable.

From this point on, I changed my environment and fed my mind with spiritual and positive things, and as a result, I'm a Professional Engineer, LEED Accredited Professional for Build Design and Construction and an Adjunct Professor. This was only the tip of the iceberg.

**Workbook Exercise**
Now that you understand why persistence is
a vital element to engineering a life of success.
Complete exercise #4 in the workbook.

Download workbook at
<u>www.rickyventers.com</u>

# *Creative/Imagination*

*Speaking is my art, and I paint pictures when I sing.*

*Visualize these musical notes;*
*I'm bringing a craft to the game.*

*No one can match what I do,*
*I'm creative when I free myself.*

*Free my mind and remove the mental boundaries*
*my skill set is my wealth.*

*One of a kind, when you see my work,*
*you can't help but be amazed.*

*Give me a platform and watch me set it fully ablaze*

*I'm on fire, watch me work.*

## —IAN FERGUSON

# CREATIVE/IMAGINATION

Creativity/Imagination [ *CI*]
The ability to transform an object within the
bounds of current limitations.

It was another one of our long drives out to a job site. My two colleagues and I made this particular trip at least twice per month, but this time was a little different. We were asked to survey an existing steam tunnel. The tunnel was close to 100 years old and not much work had been done on it in the past. With little information on the conditions of the tunnels, we brought everything we could in preparation to handle any situation, or so we thought. We spent a significant portion of the trip discussing the game plan of how we would conduct the survey. We knew who we would call to get access and, what information we would like to retrieve for analysis.

We finally arrived, and one colleague quickly realized he forgot to bring his work bag. He quickly scrambled and gathered some loose items he had and placed them into a plastic shopping bag. My other colleague and I couldn't help but laugh at the sight of him walking around with the plastic bag. This should've been our first sign of realizing we weren't as prepared as we thought for what was to come. The next sign we should've picked up on was the facilities personnel willingness to hand us the key and informing us he would not be joining us on our excursion. Still naive to what was to come, we quickly walked over to the first entry point of the tunnel and opened it up. The "door", if that's what you want to call it, was a two-foot by two-foot louver opening and creaked as we swung it open. My colleague stuck his head in first and immediately requested a flashlight. The look on his face when he pulled his head out was priceless. Now let me give you a little context here. Both of my colleagues were either at retirement age or a couple of years from it. The number of things they have seen over their years are unquantifiable, but with one look into that tunnel, he knew this tunnel experience was going to be something talked about for years to come. I had also recently recovered from an Achilles rupture, so we all knew we were going to be pushing our limits. We all managed to squeeze ourselves into this tiny entry. We stood there with flashlights in our hands, excited that we accomplished our first feat. Now it was time to get to work.

A few hours later we found ourselves completely immersed in the project. We were crawling over, and under pipes on our backs, stomachs and knees. It was crazy and a mess, dirt was everywhere. I told you this whole story just to get to this one part. As we were in the tunnel and in shock of all the old steam pipes stacked in there, I looked at one of my colleagues, and he said these words I'd never forget "Engineering is an art." "As an Engineer, the ability to take something existing in whatever state it's in and convert it into something functional takes creativity and imagination. "It's an art." At that point, my mind shifted from standing in a mess to standing in the midst of a challenge. I need to activate my creativity to find the solution. Ask yourself these questions. Do I see a mess or problem? Do I see impossibilities or an opportunity to grow?

What happened to your creativity? What happened to your imagination or dreams? One of the most dangerous things that could happen to us is that we stop searching the deepest thoughts of our mind. One day I watched the Lego Movie with my boys, and one thing they illustrated was a city where everyone did the same thing every day. Everyone had the same routine. Sang the same songs and watched the same TV shows. "Everything is Awesome…" you know the rest of the song. As you can see in the movie, it produced a mindset of limited individuality. Then came along the Master

Builders, builders that used their creativity and didn't design based on the instructions written out. Unfortunately, these master builders were considered outcasts, why? Simply put, they refused to look and act like everyone else. The leaders of the city thought that by removing creativity, they would create unity, which in some cases they did, but that form of unity would never allow individuals reach their fullest potential. They would always look to be told what to do next, never exercising the initiative to think on their own. It was a great movie, but I'm afraid that many people walked away not realizing they are falling into the same trap.

Life has become so busy for us. We wake up in the morning with just enough time to prep for work. Then we travel in our cars listening to music or the news; some people are even finishing dressing all while sipping on coffee and eating a breakfast sandwich. Then we get to work and comb through our long list of e-mails for a couple of hours before we go to our first meeting, then the second one which leads to lunch. After lunch, we try to complete an assignment that's been on the desk for a couple of days before we go to our last meeting of the day. On the way home, we sit in traffic and try to unwind our minds before pulling up to the house knowing once you get home after you pick up the kids, it's going to be homework time or baseball practice. After homework and practice, it's time for the kids to go to sleep. Now it's 8:30–9:00 p.m. depending on the day, and you finally get to sit down and rest. You watch a couple of episodes of your favorite show on Netflix before you fall asleep just to wake up to do the same thing again the next day. Sounds like the movie right "Everything is Awesome…" If we are not careful, we can go through years without realizing we are living a life that is hindering our creativity. A life in which we are very busy, but are not fruitful. A good friend once reminded me of a scripture in the Bible that states "Be fruitful and multiply." He pointed out that there is a vast difference between busyness and fruitfulness. Busyness has all the characteristics of a hard worker, but it's a hard worker with no vision. Someone that routinely does that same thing every day with no thought of why they are doing it. Fruitfulness is action taken with a mission and will always lead to positive results and add increase. The only way to create that mission is to dream, imagine and create with your mind. If you don't allow time in your day for this, you will be forever carrying out others' dreams.

Another reason why people don't use their creativity and imagination is the fear of not being accepted. All of those wild dreams you had as a kid you pushed away because at some point someone told you it was "childish." If you think about it, imagination and creativity are in its purest form when a child is playing. Anything that comes to mind, they try to act on it, but as they grow older, their influences start to shape their limitations. So I'll say to you, be that child again. Imagine and explore the deepest parts of your mind, and let innovation be birthed from you. Awaken that creative genius inside of you, and take part in impacting the world around you and beyond.

**Reflection**

When you embrace your God-given ability to be creative, it will allow you to see life in an entirely different perspective. Challenges won't seems as threatening anymore, and new ideas will birth from your mind. In other words, every challenge will bring you closer to discovering the full potential that's inside of you. That's why creativity is so important. We should be on a never-ending quest to expand our mental capacity and explore the deepest thoughts of our imagination without fear of being different.

How do we activate our creativity within?

1. **Prayer or Meditate**

   I wake up every morning and spend time reflecting on the areas in my life I need to improve on. This time is so refreshing and calm giving my mind, body and soul time to rest. My most creative thoughts have been a result of one on one time with God.

2. **Read**

   Find books, magazines or newspapers covering subjects that interest you. I remember walking into a mansion that was transformed into a museum and one wing of the house was a library. Many of us may have a bookshelf or two but imagine a house where there are books that stack on shelves from floor to the ceiling on all four walls of the room. By the way, the ceilings were 15 feet high. By reading, you will expand your knowledge, and it will spark ideas to improve upon what already exists.

### 3. Exposure

Expose yourself to new things. Travel to different areas of the world. Take up a new hobby.

### 4. Sleep

Get some rest. I can understand if you're working on a project as I am now with writing this book. This will probably result in you losing some sleep at night, but if you're up late on a continual basis just to watch your latest episode of a show, you're not doing your body nor your brain any good. You will not perform at an optimal level.

**Workbook Exercise**

Now that you understand why creativity is a vital element to engineering a life of success. Complete exercise #5 in the workbook.

Download workbook at
<u>www.rickyventers.com</u>

# *Confidence*

I face the fears and turn my uncertainties to cheer

Approaching the situation dead on, tell the challenges
ahead, "I'm here."

All with a smile, and a smirk somewhat senile

Confidence so high, anyone,
even Mr. Wonder can see it from a mile

I can't predict the outcome, but I know what I can do

I preview the view, set my standard high because my
attitude is brand new

Not to be confused with cocky, life humbled me with
loans and college tuition

But my head is held high, making the word
Focus come alive

You can see it in my eyes

As I'm speaking it into fruition,
my eyes are on the mission

My biggest challenge is myself,
this is a solo competition.

**—IAN FERGUSON**

# CONFIDENCE

A feeling of self-assurance arising from one's appreciation of one's abilities or qualities.

There are two ways to look at confidence, external or internal. External confidence is building trust, belief or faith in someone else. Internal confidence is as defined above a feeling of self-assurance arising from one's appreciation of one's abilities or qualities. For this element, I choose to talk about the internal aspect of confidence. Both are crucial and should be developed, but I realized that the first step in excelling in life is believing you can. Confidence is relative and not only based on an individual's exposure to a particular thing, but also dependent on an appreciation of their ability. Many people will never display confidence without the combination of both of these.

Take my two youngest sons, Levi and Kaleb, for example. They are thirteen months apart, and the complete opposite in many ways. We placed them both in swimming classes at the age of 4 and 3. They were both excited about the idea after seeing their older brother swimming for a couple of years. The first few days of class, they were so excited. They had their new swimming trunks and towel, and both walked in and waited on the side of the pool eager to get in, or so we thought. Their names were called, and Levi jumped right in, Kaleb on the other hand just stood there. The instructor called him a couple more times, but he turned his head to prevent eye contact. "Oh boy," I said to my wife, Nicole as she gave me the nod to go in and assess the situation. I walked in and squatted down to have eye to eye contact with my son and said "Hey bud I thought you wanted to swim" he quickly answered, "No I don't." I stared at him dumbfounded for a second thinking of what to do next. Then I immediately began rolling up my pants leg took off my socks and shoes and took him by the hand and walked him in and stood there with him for the remaining of the class. This went on for the next three weeks. I tried to give him the pep talk in the car, but no matter how excited he got in the car when he saw the water he froze up. Meanwhile, Levi was having the time of his life, splashing around and playing with toys. After that third week, we finally had a breakthrough. Kaleb was finally confident enough to get in the water alone. Towards the end of that session, things started to shift a little. Levi was very confident when it came to playing and splashing around, but when it came to dipping his head under water, he was a little hesitant. Let's fast-forward to a year later.

So during the 2016 Rio Olympics, the family would watch a few events together every evening. This was the first time the boys understood what was going on to some extent, at least. While watching the diving competition, Kaleb looked at me with excitement on his face and said "Daddy I want some swim underwear," I laughed a little and responded, "Do you mean swimming trunks?" He replied, "No, I want swim underwear."

From that day on it seemed like swimming was all he ever talked about. So after a few months break from swimming class, we started him back up. To our surprise, he had no fear what so ever and I don't mean he was just able to walk in the water by himself, but he thought he could dive. Of course, Nicole and I freaked out because we both knew he couldn't swim, but he just kept going. Let me clarify something when we finished the first swim class, the most they could is put their head under the water; they hadn't learned any strokes or kicking. So to see Kaleb jumping in and swimming away completely shocked us. Levi still loves the water, but his confidence in his ability is slowly growing.

Note I said confidence in his ability. They both have the same amount of exposure to the water, but the difference in the level of performance is dependent on the confidence level they each have in themselves. Now when we talk about legos, Levi grew a passion for building to the point he would try to create the characters he saw in a cartoon. The types of things he's able to build amazes me. As with legos too, they both have the same level of exposure, but Levi has excelled faster because of his confidence level in his ability. It's amazing to see the progression of two boys with very similar exposure points displayed different levels of confidence.

I used to use the words "I believe so" when answering a question about a project. I thought by saying this, I could somehow convince my colleague I knew what I was talking about without really being confident at all. After a few times, my partner caught on the tactics and told me "Ricky, to do well in this business you must be confident in what you know. Clients will not pay us if we are not confident in the product we are producing." I had to self-assess the reasons I wasn't confident in my final product and what I found was that my lack of confidence came from my lack of exposure and appreciation of my abilities. My mentor used to always tell me "Ricky, you have a lot of potential, but there are steps you need to take to really advance to the next level."

So what did I do? How did I overcome the lack of confidence in this area? As I said, through my self-assessment I realized one area that contributed to my lack of confidence was due to lack of exposure. Before I go on any further, I must pause and say people can have all the confidence and belief in you, but if you lack faith in yourself, you will be stuck in life. The great part about it is that you are in control of it. If you're waiting for someone else to instill internal confidence in you, you will never see. A coach can design a play for you in the game because he sees the ability that you have, but once you get the ball in your hands, it's up to you to be confident enough to carry through the game plan. So with that said, exposure was the contributing factor to my lack of confidence. I had to take it into my hands to change. I read as much as I could. I picked up every book I could read related to what I was doing. I sat down with Senior Engineers to gain deeper knowledge about different topics. I volunteered to be a part of committees to be exposed to different methods of dealing with particular designing standards. The biggest thing that helped me was asking a lot more questions. I questioned everything seeking to get clarity in any way that I could. Over time, I found myself growing in my confidence. This process did not happen overnight, but I realized that over time the company started to put more trust in me and from that, I received a promotion to Project Manager.

I interact with many business owners from small businesses to multi-million dollar companies, and one thing I've come to recognize is that one difference between the ones that are most successful and the ones that aren't, is the level of confidence the leadership has in the company's ability to perform a job. A high standard of confidence will not only propel you forward as an individual, but also place you in a position of leadership. Now leadership is a book within itself, but having confidence will contribute to becoming a great leader.

## Reflection

Many times we try to put on the appearance of confidence, but if someone could see through our facade, they would see the insecurities. Many of us work hard on the facade and making sure it's perfect and flawless without spending any time working on our internal self. I loved the type of work I did in engineering. There was nothing glamorous about it, but the lack of any system we provided would leave a building in a vulnerable state. Imagine this; you see a beautiful building on a hill. It has large windows to let the rays of the sun pierce through the building, brick exterior, and a manicured lawn. You are so excited to go inside to experience its beauty. As you walk inside you are in awe of the layout. You see the office areas, dining area, shared space and open reception area; it was a great site; however, something didn't seem right. You walked in the bathroom, and there were no fixtures. There was no toilets, sinks, water or exhaust fan, for that matter. As you continued to walk around the building, you began to notice a few other things. There were no lights in the building. Yes, the sun was shining bright, but once the darkness set in, there would be no way for you to see. You also started feeling very hot as you walked around and realized there was no air in the building and the list went on. Not only was there no water in the bathroom, there was no water anywhere in the building and in addition to there being no water there was nowhere to plug in your cell phone or any electronic device. I think you get the point. This building in all its beauty was not functional. This is what the lack of confidence is doing to you.

The facade you display is only covering the emptiness inside and not allowing you to grow. Just like the building, one day someone is going to take a peek inside and see past your exterior defense and find a person that's struggling to find security in who they are.

There are many ways to develop your confidence, but I will share with you three to get you going.

### Mindset

You have to believe that you are capable and there are no limits on your life. I believe that if the God I serve believed in me enough to send His Son to die for me, then how I could not believe in myself?

Begin every day reading books or listening to positive speeches that will encourage you to strive for excellence. One staple that I always read is the Bible, but let me also give you a list of some individuals I would suggest reading or listening to their materials as well. I recommend Eric Thomas, Jim Rohn, Malcolm Gladwell, Jeremy Anderson, Gene Bedell, Dave Ramsey, Zig Ziglar, John Maxwell, Vaughn Edmeade, Brian

Thomas, Ty Douglas, Mason West and David Shands.

### Practice, Practice, Practice

"You're talking about Practice" the famous sound bite of NBA Hall of Famer Allen Iverson. Yes, Practice! Nothing builds confidence better than repetition. The more you do something, the better you get. It's simple, but the application of it can be difficult at times. I remember playing the piano, and my parents had me practice my songs over and over and over. I didn't understand it at the time until it was show time and the nerves started to sink in. The only thing I could rely on was that my fingers have played through the song so many times it was as though once I placed them on the piano, they took it from there.

### Connect with Gurus

I cannot stress this one enough. Find people who have traveled the path that you are pursuing and allow them to give you guidance along your journey. There is nothing better than learning from other people's experiences. They will give you insight that will propel you in the right direction.

Ian and I have a podcast called The STEM Dialogue which we allow STEM students to get a glimpse into their future by speaking to professionals in their chosen major of study. The podcast has reached 31 countries 30 States and continuously grows each month.

## Workbook Exercise

Now that you understand why confidence is a vital element to engineering a life of success. Complete exercise #6 in the workbook.

Download workbook at
www.rickyventers.com

# Team Player

*A cord of 3 strands isn't easily broken,*
*that's what the Word says.*

*So why should I attack my tasks alone,*
*be Mr. Independent?*

*Instead, I have to use my resources,*
*be effective by the minute.*

*My team is like a pack of wolves, a pride of lions;*

*there's no denying this, if I did I'd be lying.*

*We all have different positions,*
*but the same goal to achieve.*

*Project after project, with my team we stand tall;*
*we don't fall like the leaves.*

*If anything, we mesh together,*
*different personalities and all;*

*God brought us all together, for a greater purpose,*
*to do what he called.*

**—IAN FERGUSON**

# TEAM PLAYER

Team Player - a person who plays or works
well as a member
of a team or group

I was young when I learned the value of teamwork right on the football field. I started playing pop warner football when I was ten years
old, and year after year a group of us would advance from level to level as we aged. We grew to
a point where at the beginning of each season,
it only took us a matter of a couple of weeks before we were in sync again. The beauty about
team sports is that the only way to be successful
is if each person on the team plays their position well. The moment you miss your route or
block you are putting your team in a vulnerable
position.

I remember being so confident at the beginning of games because I knew I had a group of
guys that were going to bring their best to the
field; all I had to do was my part. Two plays

that I remember the most was the 434 and 327 pitch pass. We called the 434 our bread and butter play; it was when all of the backs, two running backs, and a fullback ran through the same hole. Two of us blocked, and the other carried the ball. The play always seemed to work. I remember playing fullback at times and being the first one through the hole and blocking whoever was in front of me clearing the way for the other two coming behind me. There were also times where I played running back, and for the 327 pitch pass, I was given a passing route. All I said to myself was, "Get to the spot and the ball will be there." Our quarterback started his cadence "Down, Set, Hut, Hut" the ball was hiked, and I took off down the field. The sequence of the play was for the quarterback to pitch the ball to the other running back on a sweep. The running back would then stop and throw the ball down field to me on the  passing route. It was third down and five, we were on the 20-yard line. I ran my route perfect and just as I turned the ball was within reach I stretched out and caught it for the touchdown. It was an excellent display of teamwork. Everyone executed their task with perfection. There were a lot of lessons learned on that field and with the help of great fathers and coaches, we all went on to do great things in life.

Being a team player forces you to humble

yourself for the betterment of the entire team and focus on completing your role with excellence. As an engineer, I constantly have to collaborate with other engineers. When designing mechanical systems, we always need the support of electrical engineers to power our equipment. It was critical for our two trades to coordinate our drawings. The lack of coordination could cause the incorrect power supply to the buildings. Nobody would want to be the bearer of that bad news. Being a team player forces you to think beyond your bubble of thought and always keep in consideration of the scope of the project. Coordination of trades is a tremendous effort, and sometimes it would take multiple meetings where all designers would be present, providing their opinion on where things should go for everything to fit. Technology has made the process more efficient by allowing engineers to work in sync and avoiding conflicts while they are drafting.

John Maxwell says "Teamwork makes the dream work, but a vision becomes a nightmare when the leader has a big idea and a bad team."

It didn't take long for me to realize that no great thing can be accomplished alone. For example, the Yankees ruled baseball in the late 90's going into 2000's. They had players like Orlando Hernandez, Roger Clemens, Andy Pettitte, David Cone, Mariano Rivera Tino Martinez, Bernie Williams, Ricky Ledee, Scott Brosius, Derek Jeter, David Wells, Chili Davis, Paul O'Neil, Jorge Posada and Shane Spencer. They weren't entirely free of problems, but at the end of the day, they won four championships within a five year period. It was incredible! They created a culture of winning by everyone displaying teamwork when it meant the most. With all the great stories of teamwork, you also have those instances where even if a team of great players were put together, the lack of teamwork prevented them from reaching their potential. I watched in disbelief where on two occasions my favorite basketball team, LA Lakers, had a starting lineup of Hall of Fame caliber players, and to my dismay, each season ended in disappointment.

Listening to a leader at a large organization, the question was asked, how do you solve a lot of the problems that you face? He just replied, "I surround myself with gurus, and we work as a team." Teamwork is the fundamental cause of corporate success, learning what your role is, and carrying it out with excellence that places your team in a winning position.

But what's in it for me? That's one question I often hear. How do I benefit individually from being a team player? Unfortunately, we are in a selfish society, and with every passing generation, it gets worse and worse. Everyone has to get some 'shine time.' If you're not leading in the office, field, or stage, then you eventually lose interest. This isn't an attitude that you develop overnight nor is it something that you can correct overnight, but one thing I urge you to do is see the bigger picture. Like Zig Ziglar always said, "You can have everything in life you want if you will just help other people get what they want." By working together, everyone wins. I know sometimes it may feel as though you are doing so much more for the team and you often don't see the positive results, but trust me over time your teamwork will benefit you in the end.

By applying this element to my professional career, I've seen team members bend over backward to help me on projects, why? Because I've been there for them when they needed me. There have been multiple times where I stayed at work late or traveled long distances for the sake of the team, and in the end, all the support I've ever needed was available.

## Reflection

Remember these three quotes:

*"Teamwork makes the dream work, but a vision becomes a nightmare when the leader has a big dream and a bad team."*

### JOHN MAXWELL

*"You can have everything in life you want if you will just help other people get what they want."*

### ZIG ZIGLAR

*"It's amazing what you can accomplish if you don't care who gets the credit."*

### HARRY S. TRUMAN

How do you become a better team player?

1. **Understand the Objective**

   When a clear mission and vision are communicated, you can have a reference point as you progress through the project.

2. **Understand your strengths and weaknesses**

   Everyone on the team should be operating in their strengths; therefore eliminating the impact of weakness any individual may have.

3. **Seek to add value**

   Always have the goal to increase the value to whatever team you are a part of. You want the team to be better than you found it.

4. **Make sure everyone wins**

   When everyone's goals and dreams are fulfilled, you will have a group of people that are committed to helping you on future endeavors.

## Workbook Exercise

Now that you understand why being a Team Player is a vital element to engineering a life of success. Complete exercise #7 in the workbook.

Download workbook at
www.rickyventers.com

# *Authentic*

*I'd like to apologize for not being the true me.*

*Living and settling for what I thought
others wanted to see.*

*But there's only one RV, Ricky.*

*God made me in his image so when
I hide behind a false illusion*

*I can hear the Lord saying to himself
"Why won't my child let me use him?"*

*Walk with him, forgive his sin and love him as my kin.*

*Ashamed of me? It's sad to see.
I need my boy to embrace his originality.*

*And carry it out for the world to see,
he's not just my son, but also my student, I'm like a
professor at a University.*

**—IAN FERGUSON**

# AUTHENTIC

Authentic – to be completely original; genuine

One of my friends, more like a brother, called me one day and asked me if I wanted a pair of Jordan's. I was excited and said, "Absolutely yes!" Who wouldn't want a pair of Jordan's? I had never bought a pair, so this would be a first for me. A couple of weeks went by until he came by the house and just like he said he had the Jordan's in his hand. I was ecstatic and couldn't wait to wear them. We had a big event coming up, and these Jordan's would make my outfit complete.

The day came, and I slipped on the new Jordan's and was on my way. Halfway through the day, I noticed my feet started to feel uncomfortable and I wondered to myself "Does Jordan really play in these uncomfortable shoes? I would never wear these while playing basketball." By the end of the day my feet were done, I could barely walk and as I was explaining my discomfort to my brother I asked the question I should've asked when I got them "Where did you get these from?" He responded, "they were

RICKY VENTERS P.E.

shipped from China." I was wearing fakes! I'm
not the sneaker guy so there would've been no
way for me to notice they weren't the real deal,
but once I put them to work, the integrity of the
shoe was revealed.

We live in a society where finding authentic people is hard to come by. Social media has
provided so much exposure of the lives of individuals around the world, and we often try to
model our lives based on what we see. I remember reading the hashtag of an Instagram post
that read #nofilter. I laughed because clearly,
the photo had a glossy layer placed on it. The
most disturbing thing was that people com-
mented: "You look great." I once asked a group
of summer interns "Who would they be without the filters on their phones? Could they even
identify their true self? When they look into the
mirror who do they see?" Sadly many of them
could not answer the question.

There's a difference between emulating a character trait of a person versus trying to become a
split image of an individual. Character traits are
a product of the application of adopted principles to one's life. They are unique to each individual based on their life's experiences. There is
no way you can become the split image of someone without experiencing the same exact life cir-
cumstances which includes DNA make up. Just

like those Jordan's, you may look genuine, but when you're tested, you will be exposed.

Companies have seen some success developing "knock-offs," but that success only last for a limited time. People are looking for the authentic you because of the simple fact that there is no one on this earth that's like you.

If you look at many musical artists, you will see that there are microscopic similarities between the top singers. Their voices are different, their look is different, even their stage presence is different, and through their difference, they are all respected.

Finding your authentic self can be challenging. We are saturated with advertisements to become more beautiful, more fit, buy more things, become the next YouTube sensation or get a bigger and better home. It's a constant barrage be more, do more, have more, and the whole time you haven't spent any time asking yourself the question who am I and who am I supposed to be? It could be years before you realize you're living a life that's not in line with who you were destined to be. I came to this crossroads during my time as a design engineer. In college, I knew I was going to either become a doctor or engineer. I wasn't sure what type of engineer I wanted to be, but I knew it was something I was interested in and could do. After giving up on becoming a doctor due to long years in college and medical school that was associated with it, I narrowed down my engineering options down to Mechanical Engineering. I performed well in school, but while most of my classmates enjoyed many of the technical courses, I found my interest in the course call Design Optimization and Methods. It was a more of a strategic thinking course to optimize processes. I graduated and eventually found a job as a mechanical engineering consultant. I'll tell you more about this in the last chapter of the book. I had no clue what a mechanical consultant did, but over time after going through some rough patches, I became very good at what I did. But there was still something that left me unfulfilled. It wasn't until I was promoted to project manager that my passion started to come alive. It was exactly where I felt I could thrive. Don't get me wrong; I want you to fully understand that most times you have to go through levels in your career or life in general just to get to where you want to be. I truly appreciate the process I had to go through, but I'm most grateful I was able to find a place where I could be my authentic self. It was in that place that I began to turn my striving into thriving and grow exponentially in my career.

## Reflection

Sometimes when you want to find your authentic self, you need to remove all distractions from your life. What is it worth to you? Are you willing to separate yourself from social media or TV in general just to take time to soul search? I know I sound like an extremist, but no one is going to take your life more serious than yourself. So if you're waiting for someone to hold your hand and do all the work for you, you will be waiting for a long time, if not forever. I had to get to the point in life where nothing else mattered more than for me to find who Ricky is and what am I called to do. You will never fulfill your purpose in life trying to be something or someone you're not. You can't fake your way through life. Yes, it may work for a while, but just like those Jordan's, you will never pass the test of time. Time will reveal if you're genuine or a knockoff. Take the time now to identify who you are and grow in confidence of being your authentic self.

## Workbook Exercise

Now that you understand why being Authentic is a vital element to engineering a life of success. Complete exercise #8 in the workbook.

Download workbook at
www.rickyventers.com

# CONCLUSION

Though we may have difference experiences, the application of these eight elements in your life will produce a successful life. There are going to be many moments in life where your progress may seem minuscule compared to your coworkers or friends, but I encourage you to stay laser focused on your goal. What is your goal? Who are you destined to be? These two questions will either haunt you or motivate you. Through my years of education, I only focused on the moment and never actually took the time to look at the big picture of life. I was in school to become a mechanical engineer, but honestly, outside of my course work, I didn't know what it all entailed. So when moments got rough, it became very challenging to cope. Listening to my father's advice, I never gave up, but there were many times that I felt a strong desire to quit. As I look back over my life, I can see how every phase and experience built upon the next. The lack of these experiences would have been detrimental to my growth.

I believe everyone is given a divine purpose that will impact the world around them in their way, but this blessing could lay dormant for years if the proper steps aren't taken to activate it. As God started to give me a glimpse of what my future had in store, I realized that I wasn't just being called to be a mechanical engineer. I was called to be an Engineer of Impact. I am called to change lives around the world sharing my journey of how with hard work and determination you can make your dreams become a reality.

How did I go from being a math failure to where I am now? The first thing I had to do was self-asses my performance in school. It wasn't that I was incapable, it was merely a lack of effort on my part. For a long time, I blamed the teacher for telling me I would never succeed until I realized no matter how much blame I put on her it wouldn't make me any better. One big move that propelled me in the right direction was when my mother started to homeschool my siblings and me. I remember people asking, "Were you a troublemaker at school? Why would your parents do that?" or saying "Homeschooling isn't going to be good for them. They will struggle to fit in socially." It was funny because everyone had a preconceived notion of what our outcome would be. To their surprise when my brother and I got a 4.0 GPA and 3.78 GPA respectively our first semester in community college and were offered opportunities to tutor chemistry and math those negative perceptions were challenged. Yeah, my brother got a higher GPA, and I failed to mention my brother was 14, and I was 16 years old. No one on earth has pushed me harder than my younger brother, Ravon. Sports, academics, whatever you name it our relationship and level of accountability to each other brought the best out of me.

I went through my college years first getting my associates in liberal arts and science, then went to Stony Brook University and majored in Mechanical Engineering. Those were a rough three years, but by creating a positive environment, I was able to survive. One thing I didn't take advantage of during undergrad was internships. Instead, I devoted my time as a mentor for a College Prep program called Upward Bound every summer.

After my undergraduate studies, I hit a rough patch. I foolishly turned down the only engineering job offer I received, and I couldn't find another one for an entire year. People said I was foolish for not doing an internship and I felt I just wasted my time away every summer. During that year I enrolled into graduate school not sure what to major in so I just stuck with mechanical engineering. I managed through the help of a church friend become a substitute teacher who helped put some traveling money in my pocket. My days began at 4:30 am and ended at 12:30 am. Yes, that was not a typo you read that correct I left early in the morning to teach and got back home from my night classes after midnight. My stress levels were extremely high I felt like a failure and felt my life was going nowhere.

I felt my relationship with God was broken, and my prayers were never going to be answered. One day during my worship and meditation I came across a scripture from the Bible. "You are good, and what you do is good; teach me your decrees" (Psalm 119:68). I know everyone isn't religious, but I would do you a disservice to share all things that happened in my life without sharing the source. When I read this, I began to cry because at that point I realized nothing that happened to me was by chance and that it will be good in the end. So, from that point on every single day and night traveling, I would repeat "You Are Good." This is what got me out of my dark times and allowed me to refocus my energy on how I can take advantage of the time I was given.

At the end of the school year, I went back to working with Upward Bound, but something inside of me told me this would be my last year serving in this capacity. While going through my e-mails one early morning I came across a correspondence with the company which I turned down the year prior. I decided to reach out to them and share my experience over the past year. Nothing more or nothing less. Later that day, I received an e-mail from the company about an opportunity that opened up and that they would like for me to come in and interview. I was extremely excited. I wasn't expecting to hear back from them let alone be offered another interview. I went to the interview, and before I reached home, an offer was in my inbox. In one day it felt as though my entire life changed. Everything I dreamed of and prayed for was starting to take shape. Shortly after I got married to my beautiful girlfriend of 6 years, Nicole, we had three healthy boys Solomon, Levi, and Kaleb. Life was great.

As life would have it, my family and I relocated from New York to Maryland. Everything was new. New job, house, neighborhood, friends and church, but one thing that didn't change was applying the eight elements I've adapted. I've learned that these life principles, when applied, will produce the same results over and over no matter the circumstance.

It was here in Maryland that the fruit of all my labor started to take shape. While working as a mechanical engineer my engineering and leadership skills grew. It was at that time

I began to feel a strong desire to start speaking and mentoring students in middle school, high school, and college. The inspiration and motivation I was able to share with them sparked a fire inside of me to do even more.

As I grew in engineering, I noticed my interest started to shift a little to Construction Management where it would allow me to do more project management. After about two years in Maryland, a great opportunity opened up for me to work with the #1 Hospital in the World, Johns Hopkins.

It couldn't have been a better time. While at Johns Hopkins I was encouraged to find a chance to give back to the community. I meet three gentlemen named Maurice Wagoner, Ian Ferguson and Doron Eisenberg and together we formed a team that would create a Non-Profit Organization called InVenters. Also during this time, I received my Professional Engineering License (P.E.). In the words of Randal E Reibel, P.E. "Basically being a PE means that you are at the top of your game, top of your profession. They don't just hand that out to anybody." As reported by U.S. News only 20% of engineers are professionally licensed and further researched showed that of the 20% only approximately 2% are minorities. This was a huge accomplishment for me, but even with this success, one thing I realized was that I needed to help expose more students to this industry and encourage them through the difficult times it will present.

I warned you that when applying these eight elements to your life doors will open you never imagined. At the same time, I started to speak to students roughly four years ago. Then Johns Hopkins began to discuss how they could make a bigger impact in the community. Shortly after I received my P.E., a leadership opportunity at Johns Hopkins opened up for me to step into a role that requires my mentoring, engineering, construction management, speaking and workforce development experience. Something I could've never planned in a million years.

Everything came in full circle. Every phase of my life has added value to where I am today. In retrospect, I see there was no time wasted. Yes, even those nights when I was traveling home late and confused about what direction my life was going in. My experiences were all pieces to my puzzle of life. I was reading a story to my boys one night and a statement resonated with me. "Life is like a puzzle, even though a single piece may seem insignificant without it the beautiful picture would be incomplete."

Just like you, my story is still being formed, but one thing I know is that none of this would've happened if I didn't submit to my

God-given purpose in life and apply these eight key elements that shaped my character to what it is today.

To be continued…

THE STEM
DIALOGUE

Ricky Venters P.E.

Ian Ferguson

Podcast Available on
iTunes

Available on
Podbean